ume 30 No. 1 2000 ISSN 0146-6453

Annals of the ICRP

PUBLICATION 84

Pregnancy and Medical Radiation

Annals of the ICRP

ICRP PUBLICATION 84

Pregnancy and Medical Radiation

Editor
J. VALENTIN

PUBLISHED FOR

The International Commission on Radiological Protection

UK	Elsevier Science Ltd, The Boulevard, Langford Lane, Kidlington, Oxford OX5 1GB, UK
USA	Elsevier Science Inc., 660 White Plains Road, Tarrytown, New York 10591-5153, USA
JAPAN	Elsevier Science Japan, Tsunashima Building Annex, 3-20-12 Yushima, Bunkyo-ku, Tokyo 113, Japan

---

First edition 2000
ISBN 008 043 9012
ISBN 13: 978 0 08 043901 3
ISSN 0146-6453

Published quarterly (March, June, September, December)
March 2000 issue

Typeset by Variorum Publishing Ltd, Rugby
Printed and bound by CPI Group (UK) Ltd, Croydon, CR0 4YY

Annals of the ICRP

ICRP Publication 84

Editorial

BACK TO BASICS

Sometimes it is useful to sit back and recall a bit of history. To the layman, protection against ionising radiation (if at all known to be something else than non-ionising radiation) is usually taken to be concerned with the nuclear fuel cycle. Yet the biggest man-made source of exposures is not at all nuclear power but medical uses of radiation. Radiation in medicine is a wonderful tool that has permitted amazing improvements in diagnostic capability, great curing capacity, and invaluable palliative treatments. That same tool can also be lethal unless treated with the utmost respect — and that is where ICRP, the International Commission on Radiological Protection, started.

The huge medical benefits of radiation were recognised almost immediately after the discovery of ionising radiation at the end of the 19th century. The dangers of radiation soon became increasingly apparent, and in 1928, what is now ICRP was established by the 2nd International Congress of Radiology under the name of the International X-Ray and Radium Protection Committee. Thus, it was the medical community that first saw the need for a specific international organisation devoted to protection and, to this day, the Commission retains a special relationship with the International Society of Radiology (the professional organisation catering for physicians specialising in the use of radiation).

Back in 1928, the primary concern was that occupational exposure of medical staff could, and did, lead to serious deterministic harm and even death. With improved standards, this hazard seemed to be virtually eliminated several decades ago. Increasing awareness about possible late stochastic effects necessitated a refocusing onto the risk of cancer and onto patients rather than staff.

At much the same time, increasing use of radiation in other contexts than medicine, and increasing attention to natural sources of radiation, were the main reasons why the Commission was re-organised and given its present name in 1950. Although one of the standing committees of ICRP is specifically devoted to radiation in medicine, the majority of the reports in the *Annals of the ICRP* deal with matters other than medical exposures.

Unfortunately, it has turned out that neither occupational hazards for medical staff nor deterministic harm from exposure to radiation are problems entirely of the past. Medical science is taking great strides all the time, and more and more subjects are competing for time on the syllabus of all medical students.

Regrettably, this has often led to absurdly short or entirely missing education in radiological protection. As a consequence, really serious skin damage and worse is again being experienced, primarily in patients but also in staff. The situation may be aggravated if inexperienced staff encounter modern fluoroscopic equipment with easily operated, potent boost options that, if used indiscriminately, produces images of a clarity which may appear seductively 'good' but is not always clinically required.

There are also a number of other problem areas, among them accidents in radiotherapy, high patient doses in computed tomography, doses to family and public from radiopharmaceuticals used for therapy, and the topic of the present report, pregnant patients and staff exposed to medical radiation.

The Commission is planning to address all of these problems in forthcoming reports. In so doing, it will highlight specific situations, often drawing on existing information and reports, but organising the material in a different, more context-oriented way than the Commission's existing reports on medical radiation which are laid out according to a logical hierarchy.

Furthermore, the Commission is keen to reach such 'shop-floor' medical staff who are involved in the day-to-day management of radiation. As a means towards this end, the style of this report, and the planned reports just described, is somewhat different than that of recent reports in the *Annals of the ICRP* (which has unfortunately come to be perceived by many physicians as aiming at health physicists, administrators, and regulators, but not at themselves). Wide-spread distribution will also be given major consideration.

This is not to say that the Commission plans to abandon its traditional type of report. Comprehensive scientific treatises will continue to be a mainstay in the Commission's series of publications. However, the impact will be augmented by also supplying topical reports aimed at special user groups, in this case medical staff. In so doing, the Commission is targeting its classical, original audience, and using a style reminiscent of that in its earliest publications. In short, with this report and forthcoming companion reports on medical radiation problems, ICRP is going back to basics.

JACK VALENTIN

Contents

Preface

Over the years, the International Commission on Radiological Protection (ICRP), referred to below as 'the Commission', has issued many reports providing advice on radiological protection and safety in medicine, most recently its *Publication 73* which is a general overview of this area. These reports summarise the general principles of radiation protection and provide advice on the application of those principles to the various uses of ionising radiation in medicine and biomedical research.

Most of these reports are of a general nature, and the Commission wishes to address some specific situations where difficulties have been observed. It is desirable that reports on such problem areas be written in a style which is accessible to those who may be directly concerned in their daily work, and that every effort is taken to ensure wide circulation of such reports.

A first step in this direction was taken at the Commission's meeting in Oxford, United Kingdom, in September 1997. At that time, on the recommendation of ICRP Committee 3, the Commission established a Task Group to produce this report on pregnancy and medical radiation. The purpose was to cover the basic issues of pregnancy and ionising radiation in medicine in a concise report, including and coherent with the Commission's current recommendations, that could be easily understood by the medical community and easily translated. The report should cover the most commonly asked questions, discuss the management of pregnant patients as well as pregnant workers, and provide a practical approach that can be used in varying situations.

The membership of the Task Group was as follows:

F.A. Mettler, Jr. (Chairman)	R.L. Brent	C. Streffer
L. Wagner		

Corresponding members were:

M. Berry	S.-Q. He	T. Kusama

The membership of Committee 3 during the period of preparation of this report was:

F.A. Mettler, Jr. (Chairman)	J.-M. Cosset	M.J. Guiberteau
L.K. Harding (Secretary)	J. Liniecki (Vice-Chairman)	S. Mattsson
H. Nakamura	P. Ortiz-Lopez	L.V. Pinillos-Ashton
M.M. Rehani	H. Ringertz	M. Rosenstein
Y. Sasaki	C. Sharp	W. Yin
W.Y. Ussov		

This report aims to serve the purposes described above. Since the report is intended for a very wide audience, its style differs in some respects from the usual style of the Commission's publications in the *Annals of the ICRP*. For instance, literature references are not given along the lines common in scientific reports; instead, a bibliography including the source material as well as useful further reading is supplied at the end.

The report was approved for publication by the Commission through postal ballot in November 1999.

Annals of the ICRP

Pregnancy and medical radiation

ICRP Publication 84

Approved by the Commission in November 1999

Abstract-Thousands of pregnant patients and radiation workers are exposed to ionising radiation each year. Lack of knowledge is responsible for great anxiety and probably unnecessary termination of pregnancies. For many patients, the exposure is appropriate, while for others the exposure may be inappropriate, placing the unborn child at increased risk.

Prenatal doses from most properly done diagnostic procedures present no measurably increased risk of prenatal death, malformation, or impairment of mental development over the background incidence of these entities. Higher doses, such as those involved in therapeutic procedures, can result in significant fetal harm.

The pregnant patient or worker has a right to know the magnitude and type of potential radiation effects that might result from in utero exposure. Almost always, if a diagnostic radiology examination is medically indicated, the risk to the mother of not doing the procedure is greater than is the risk of potential harm to the fetus. Most nuclear medicine procedures do not cause large fetal doses. However, some radiopharmaceuticals that are used in nuclear medicine can pose significant fetal risks.

It is important to ascertain whether a female patient is pregnant prior to radiotherapy. In pregnant patients, cancers that are remote from the pelvis usually can be heated with radiotherapy. This however requires careful planning. Cancers in the pelvis cannot be adequately treated during pregnancy without severe or lethal consequences for the fetus.

The basis for the control of the occupational exposure of women who are not pregnant is the same as that for men. However, if a woman is, or may be, pregnant, additional controls have to be considered to protect the unborn child. In many countries, radiation exposure of pregnant females in biomedical research is not specifically prohibited. However, their involvement in such research is very rare and should be discouraged.

Termination of pregnancy is an individual decision affected by many factors. Fetal doses below 100 mGy should not be considered a reason for terminating a pregnancy. At fetal doses above this level, informed decisions should be made based upon individual circumstances.

Keywords: Pregnancy; Radiation; Protection; Fetal dose; Termination

1. INTRODUCTION

(1) Thousands of pregnant patients and radiation workers are exposed to ionising radiation each year. Lack of knowledge is responsible for great anxiety and probably unnecessary termination of many pregnancies. For many patients, the exposure is appropriate, while for others the exposure may be inappropriate, placing the unborn child at an unjustified increased risk.

(2) One of the most commonly asked questions in relation to the use of ionising radiation in medicine concerns the management of the pregnant patient or worker. Instinctively, one might want to avoid use of radiation with a pregnant patient; however, there are a number of situations in which the use of radiation for diagnosis or therapy is appropriate. In addition to patient irradiation, there are many female physicians and technicians who are employed in medical settings involving radiation and they often wish to begin a family while remaining employed.

(3) This document is written primarily for physicians, but it will also be useful for medical and health physicists, radiation protection staff, nurses, technicians, and administrators. The document is not intended as a complete scientific reference work, nor as a collection of inflexible recommendations, but rather to provide a practical approach that can be used in varying situations. A bibliography is provided at the end of the document, should the reader wish to obtain additional information. This document will not consider exposures to diagnostic ultrasound or magnetic resonance imaging.

(4) Management of the pregnant female in a radiation setting takes several forms. If the patient or worker declares her pregnancy or is obviously pregnant, appropriate measures may be instituted. The situation is much more difficult when the female is not sure whether she is pregnant or is unaware of her pregnancy. The degree of concern, and whether informed consent for a procedure should be obtained, are dependent upon the degree of potential risk to the fetus and the mother. The potential risk to the embryo/fetus can vary widely depending on whether diagnosis or therapy is being contemplated. These issues will be covered in more detail in each section of the report.

(5) The International Commission on Radiological Protection has published a wide variety of recommendations in the past. These include general recommendations on radiological protection as well as advice on the application of radiological protection in medicine. In these and other Commission publications there has been information on pregnancy and radiation. This document will provide much of that information in condensed form and offer examples of its application.

1.1. Fundamental concepts

(6) In their daily practice, physicians and the medical community already use many of the concepts that form the basis for radiological protection.

(7) Each use of radiation in medicine should be justified (provide more benefit than harm). For example, radiotherapy is used because it has been shown that it can

reduce morbidity and mortality from cancer. After a type of examination or therapy has been justified generally, each specific instance should be justified. As an example, a standard radiotherapy protocol may be justified in a 50-year-old female but the same protocol may not be justified in a pregnant 25-year-old without more consideration and perhaps modification.

(8) Justification in medical practice is different from justification in most other practices that involve radiation. In almost all medical applications of radiation, the benefit and potential risks are to the same person. In many other radiation practices (such as the use of nuclear power) the potential benefit and potential harm may accrue to different population groups.

(9) Compared to routine medical radiation practices, medical exposure of a pregnant patient has additional ethical considerations. In evaluation of risks and benefits from medical exposures during pregnancy, at least two individuals need to be considered. The mother may receive direct benefit while the fetus may be exposed without direct benefit. On the other hand, if the mother's medical problem is life threatening, medical irradiation of the mother may lead to her survival, which obviously directly benefits the fetus.

(10) When it has been decided that a medical procedure is justified, the procedure should be optimised. This means that the conditions should achieve the clinical purpose with the appropriate dose. Too low a radiation dose will result in poor medical results, both in diagnosis and in therapy. In diagnosis the image will be too poor for diagnosis, while in radiotherapy the tumour will not be cured. Too high a dose in diagnosis increases the risk of a neoplasm, and too high a dose in radiotherapy can also result in serious and potentially fatal complications.

(11) Reducing radiation dose after a procedure has been performed is only rarely possible. One can envision a few situations in nuclear medicine in which the radiation dose can be reduced after the examination. For example, a patient may be asked to drink fluids and void after a bone scan in order to reduce bladder dose. For diagnostic radiology and most radiotherapy no intervention can be done to reduce dose after the procedure. The simple concepts of justification and optimisation will form the basis for the philosophy applied throughout much of this document.

(12) For purposes of this report, the radiation dose of interest is the absorbed dose to the conceptus (embryo or fetus) and not to the mother. Absorbed radiation doses to the conceptus are properly expressed in gray (Gy) or milligray (mGy). The gray is equal to 100 rad. One Gy equals 1,000 mGy. The unit of equivalent dose, and effective dose, is the sievert (Sv). The sievert is equal to 100 rem. Dose limits are given in Sv. In most medical decision-making applications, using X rays, gamma rays, or electrons, the numerical value of the absorbed dose in Gy is essentially equal to the numerical value of the equivalent dose in Sv. Given the homogeneity of most fetal doses in medical applications and for ease of understanding, in this document doses will only be given in Gy or mGy.

(13) For occupational and public exposure, there are recommended annual dose limits. These limits are intended to provide protection for the radiation worker and the public for sources such as nuclear power production. Dose limits do not apply for radiation exposure of patients, since the decision to use radiation is justified

depending upon the individual patient situation. As an example, a certain radiation dose may not be justified for medical screening, but it may well be justified for a very ill patient.

(14) This document begins with short sections on the diagnosis of pregnancy, informed consent, and radiation effects on the embryo/fetus. Additional sections are concerned with the management of pregnant patients who are undergoing diagnostic radiology, nuclear medicine, or radiotherapy. Sections on occupational exposure of pregnant workers and on research involving pregnant females follow. At the end of the document, there is a section on consideration of termination of pregnancy after radiation exposure, and finally a concluding section with recommendations.

2. DIAGNOSIS OF PREGNANCY

(15) **Amenorrhea occurring in a regularly menstruating woman should be considered due to pregnancy, until proven otherwise.**

(16) The diagnosis of pregnancy can be challenging, especially in the early weeks of amenorrhea. For the most part, pregnancy is diagnosed on clinical grounds without resorting to laboratory or imaging tests. Signs of pregnancy are divided into presumptive, probable, and positive. The presumptive signs include pigmentation and discoloration of skin and mucous membranes. Probable signs are related to detectable physical changes in the size, shape, and consistency of the uterus. Positive signs include detection of physical heartbeat and recognition of fetal movements. Unfortunately, most of the above information related to the signs of pregnancy is not immediately or easily available when a patient presents for an examination in a radiology or nuclear medicine department.

(17) Menstrual history may or may not be reliable in determination of pregnancy. For example, a young girl who comes to the hospital with her parents may deny a pregnancy that she suspects. Urine and serum tests to detect early pregnancy have become widely available in developed countries and are quite sensitive and reliable, and are usually positive 10 days post conception or 24 days after the first day of the last menstrual period (if the menstrual period is regular). Such tests are not generally used for low dose diagnostic examinations, but they are useful in specific instances where high radiation doses are expected to be delivered to the pelvis.

(18) In order to avoid radiation exposure in the first 2 weeks post conception, some authors have suggested limiting non-essential examinations to the first 10 days of the menstrual cycle. In most situations, this has not proven to be necessary, based upon the radiobiological and dosimetric considerations that are covered in the next section of this report.

(19) Obstetricians usually express the stage of pregnancy as menstrual age/gestational age. The latter term is usually given in reports of ultrasound examinations done for fetal dating. The two terms used in this document will either be gestational age, based upon the first day of the last menstrual period, or post conception age. Effects of radiation exposure on the embryo and fetus are expressed as a function of gestational age. Development of the unborn child is expressed as post-conception age and can be approximately divided into three major phases. These include 1) the pre-implantation phase from conception to implantation, 2) the phase of major organogenesis which extends from the 3rd to approximately the 8th week post-conception, and 3) the phase of fetal development lasting from 9 weeks until birth (which includes the important period of central nervous system development from the 8th to 25th weeks).

3. EFFECTS OF IN UTERO IRRADIATION

3.1. General background

(20) **Prenatal doses from most properly done diagnostic procedures present no measurable increased risk of prenatal death, malformation, or impairment of mental development over the background incidence of these entities. Higher doses, such as those involved in therapeutic procedures, can result in significant fetal harm.**

(21) **There are radiation-related risks throughout pregnancy, which are related to the stage of pregnancy and the fetal absorbed dose. Radiation risks are most significant during organogenesis and the early fetal period, somewhat less in the second trimester, and least in the third trimester.**

(22) Both cell killing effects and induction of neoplasms may occur as a result of in utero irradiation. Clinical radiation effects are due either to (1) cell killing or (2) unrepaired/misrepaired DNA damage. Effects due to cell killing have a practical threshold below which the effect is not seen. The higher the dose above that threshold, the more severe is the effect. Leukaemia, cancer, and potential hereditary effects are due to unrepaired or misrepaired DNA damage. The probability of such an effect occurring increases with dose and there is no identifiable threshold dose below which the chance is known to be zero. Protracted radiation exposures may occur during pregnancy. In general, for a given period of gestation, protracted exposures probably have less overall effect than a brief radiation exposure of high intensity.

(23) Above a practical threshold, damage from ionising radiation during pregnancy that results in cell killing can cause a wide range of effects, including lethality, central nervous system abnormalities, cataracts, growth retardation, malformations, and even behavioural disorders. Since the fetal neural system is most sensitive and has the longest period of development, radiation-induced abnormalities are rarely seen in humans without neuropathology. This syndrome is recognisable but can also be produced by other noxious agents.

(24) The effects of exposure to radiation on the conceptus depend on the time of exposure relative to conception and the amount of absorbed dose. When the number of cells in the conceptus is small and their nature is not yet specialised, the effect of damage to these cells is most likely to take the form of failure to implant or of an undetectable death of the conceptus; malformations are unlikely or very rare. Exposure of the embryo in the first two weeks following conception is not likely to result in malformation or fetal death, despite the fact that the central nervous system and the heart are beginning to develop in the third week. During the rest of the period of major organogenesis, conventionally taken to be from the third week after conception, malformations may be caused especially in the organs under development at time of exposure.

(25) These effects have a threshold of 100-200 mGy or higher. This dose is higher than what is reached in most diagnostic radiology or diagnostic nuclear medicine procedures. As an example, a fetal dose of 100 mGy would not likely be reached with 3 pelvic computed radiography (CT) scans, nor with 20 conventional diagnostic

x-rays of the abdomen or pelvis. At 100–200 mGy, the risk of malformations is low, but the risk does increase with increasing dose.

3.2. Effects on the central nervous system

(26) **During the period of 8–25 weeks post conception, the central nervous system (CNS) is particularly sensitive to radiation. Fetal doses in excess of about 100 mGy may result in a verifiable decrease of IQ. During the same time, fetal doses in the range of 1,000 mGy (1 Gy) result in a high probability of severe mental retardation. The sensitivity is highest 8–15 weeks post conception. The CNS is less sensitive to these effects at 16–25 weeks of gestational age and rather resistant after that.**

(27) Radiation effects on the developing central nervous system are probably the result of cell killing and of changes in cellular differentiation and neuronal migration. Values of intelligence quotient (IQ) lower than expected have been reported in some children exposed in utero at Hiroshima and Nagasaki. There have been two principal quantitative findings. The first one is a reduction of IQ with increasing dose. This effect is very dependent on fetal age. Regardless of the time of gestation, IQ reduction cannot be clinically identified at fetal doses of less than 100 mGy. In the period from 8 weeks to 15 weeks after conception a fetal dose of 1,000 mGy (1 Gy) reduces IQ by about 30 points. A similar, but smaller, shift is detectable following exposure in the period from 16–25 weeks.

(28) The second finding is of a dose-related increase in the frequency of children classified as 'severely retarded'. This is not unexpected. If the fetal radiation dose is high and there is a large reduction in IQ, there will be more children born who are severely mentally retarded. At fetal doses of 1,000 mGy during 8–15 weeks gestational age, the probability of this effect is about 40%. The effects of all levels of dose are less marked following exposure in the period from 16 weeks to 25 weeks after conception, and these effects have not at all been observed for other periods.

(29) All the clinical observations on significant IQ reduction and severe mental retardation relate to fetal doses of about 500 mGy and above and at high dose rates. Direct use of the observations for risk estimation in relation to chronic exposure of workers probably overestimates the risks.

(30) It is important to relate the magnitude of radiation effects to those abnormalities that occur spontaneously in the population, in the absence of other radiation exposure than natural background. The normal incidence of mental retardation in the population depends on the definition of mental retardation that is used. At the present time, most organisations define an IQ below 70 as mental retardation. Current prevalence figures indicate that the 'normal' incidence of persons with an IQ below 70 is approximately 3 percent. In other words, in the absence of exposure to non-background radiation, 3 out of 100 pregnancies will result in delivery of a child with mental retardation. Severe mental retardation (in which affected individuals are unable to care for themselves) occurs spontaneously in about 1 in 200 births (0.5%).

(31) There are many modifying factors. At the present time, over 250 causes of mental retardation have been identified, including malnutrition, lead poisoning,

rubella infections during pregnancy, and maternal alcoholism. At fetal doses of 100 mGy, the spontaneous incidence of mental retardation is much larger than a potential radiation effect on IQ reduction. On the other hand, at fetal doses of 1,000 mGy during 8–15 weeks post-conception, the probability of a radiation-induced significant decrease in IQ and resultant mental retardation rises to about 40%, which is much higher than the spontaneous rate of about 3%.

(32) It should be noted that radiation-induced mental retardation may sometimes be distinguished from other forms of retardation. Heterotopic gray matter and microcephaly suggest radiation or maternal alcoholism as a potential cause whereas a child with cerebral palsy, normal head size, and a documented hypoxic episode during delivery would not have irradiation as a likely etiology.

3.3. Risk of leukaemia and childhood cancer

(33) **Radiation has been shown to cause leukaemia and many types of cancer in both adults and children. Throughout most of pregnancy, the embryo/fetus is assumed to be at about the same risk for potential carcinogenic effects of radiation as are children.**

(34) As a result of radiation exposure, after conception and until delivery there is felt to be an increased risk of childhood cancer and leukaemia. The spontaneous incidence of childhood cancer and leukaemia from ages 0–15, without radiation exposure above natural background, is about 2–3 per 1000. The magnitude of risk following low-dose radiation exposure and whether the risk changes throughout pregnancy has been the subject of many publications, yet interpretation of the data remains open to debate.

(35) At low doses, the associated low risk is difficult to detect clearly in human studies. One type of epidemiological study (case-control) has shown raised risks of childhood cancer and leukaemia associated with obstetric x-ray examinations of pregnant women. Similar results have not been found in cohort studies, another type of epidemiological study.

(36) There is some evidence of a possibly raised rate of leukaemia in the atomic bomb survivors who were irradiated in utero (a cohort study with higher average doses than for obstetric X rays), but there is no increasing leukaemia trend with increasing dose and the cases did not occur during childhood.

(37) Risk can be expressed in several ways, including as relative risk or absolute risk. Relative risk indicates the risk as a function of the 'background' cancer risk. A relative risk of 1.0 indicates that there is no effect of irradiation, whereas a relative risk of 1.5 for a given dose indicates that the radiation is associated with a 50% increase in cancer above background rates. Absolute risk estimates simply indicate the excess number of cancer cases expected in a population due to a certain radiation dose.

(38) A recent analysis of many of the epidemiological studies conducted on prenatal X-ray and childhood cancer are consistent with a relative risk of 1.4 (a 40% increase over the background risk) following a fetal dose of about 10 mGy. The best methodological studies, however, suggest that the risk is probably lower than this.

Even if the relative risk were as high as 1.4, the individual probability of childhood cancer after in utero irradiation would be very low (about 0.3–0.4%) since the background incidence of childhood cancer is so low (about 0.2–0.3%).

(39) Recent absolute risk estimates for cancer risk from ages 0–15 after in- utero irradiation have been estimated to be in the range of 600 per 10,000 persons each exposed to 1,000 mGy (or 0.06% per 10 mGy). This is essentially equivalent to a risk of 1 cancer death per 1,700 children exposed in utero to 10 mGy.

(40) Excess cancers as a result of in- utero exposure have not clearly been demonstrated among Japanese atomic bomb survivor studies even though the population has been followed for about 50 years, but the number exposed is not large.

3.4. Pre-conception irradiation

(41) **Pre-conception irradiation of either parent's gonads has not been shown to result in increased cancer or malformations in the children.**

(42) As a result of early insect and other animal research, hereditary effects historically were assumed to be the major source of potential radiation harm. However, over the last three decades, risks of transmitting radiation-acquired abnormalities to offspring from irradiation of the parents' gonads prior to conception have not been identified. Comprehensive studies of the children and grandchildren of the atomic bomb survivors have not identified any heritable effects that would be linked to parental radiation exposure. New studies of survivors of childhood cancer treated with radiation therapy also have not shown genetic effects in their offspring.

(43) Nevertheless, there have been recommendations that women should refrain from becoming pregnant for several months after radiation therapy. Such recommendations were based upon experiments in mice that demonstrated that mature oocytes were more radiosensitive than immature oocytes. The use of a particular number of months for humans to refrain from getting pregnant is arbitrary. To be conservative, in the absence of a significant amount human data with doses in excess of 500 mGy, some authors still recommend that if a female receives pre-conception ovarian doses of over 500 mGy, pregnancy be delayed for at least 2 months.

(44) Most of this is a theoretical discussion of little import in practical terms. Most patients receiving doses this high usually have either significant endocrine dysfunction or cancer, and they are often asked to delay possible pregnancy for non-radiation related issues (particularly to see if there is disease recurrence). This issue is discussed further in the later section on radiotherapy.

4. INFORMED CONSENT AND UNDERSTANDING

(45) **The pregnant patient or worker has a right to know the magnitude and type of potential radiation effects that might result from in-utero exposure.**

(46) There are standard ways of looking at how to ethically justify the imposition of risk. There are three major factors: consent, role-related responsibility, and remedy/compensation. The first two are the most important for consideration of radiation exposure and the pregnant female.

(47) It may be ethically justified to place individuals at risk if they consent to that imposition. In the medical setting, the doctrine of informed consent typically requires that persons agree in advance to undergo non-emergency medical procedures after a full disclosure of the risks that these procedures may have on their health or livelihood.

(48) The level of risk that requires informed consent, and how that consent is actually obtained, is governed differently by the legal systems in various countries. There are usually five basic elements to informed consent, which includes whether one is competent to act, receives a thorough disclosure, comprehends the disclosure, acts voluntarily, and consents to the intervention. The need and degree of disclosure is usually measured by what a reasonable person believes is material to the mother's decision to be exposed to radiation.

(49) In situations involving a patient or worker, who is known or suspected to be pregnant, the situation includes not only the risk to the mother but to the foetus as well. In this setting, the mother has a role-related responsibility to care for her unborn child as well as to make decisions about herself.

(50) The level and degree of disclosure should be related to the level of risk. For low dose procedures such as a chest X-ray, the only information that may be needed is a verbal assurance that the risk is judged to be extremely low. When foetal doses are above 1 mGy and above, usually more detailed explanation is given.

(51) The information should not only include potential radiation risks but also potential alternative modalities as well as the risk of harm from not having the medical procedure. The degree of documentation of such explanations and consent is variable but many physicians will include a note of any such counselling or consent in the record of the patient or worker.

5. DIAGNOSTIC RADIOLOGY

(52) **Almost always, if a diagnostic radiology examination is medically indicated, the risk to the mother of not doing the procedure is greater than is the risk of potential harm to the fetus. Radiation doses resulting from most diagnostic procedures present no substantial risk of causing fetal death, malformation, or impairment of mental development. If the fetus is in the direct beam, the procedure often can, and should be, tailored to reduce fetal dose.**

5.1. Before irradiation

(53) **Before X-ray examination, it should be determined whether a patient is, or may be, pregnant, whether the fetus will be in the direct beam, and whether the procedure is relatively high-dose.**

(54) It is prudent to consider as pregnant any woman of reproductive age presenting herself for an X-ray examination at a time when a menstrual period is overdue, or missed, unless there is information that precludes a pregnancy (e.g. hysterectomy or tubal ligation). In addition, every woman of reproductive age should be asked if she is, or could be, pregnant.

(55) In order to minimise the frequency of unintentional radiation exposures of the embryo and fetus, advisory notices should be posted at several places within diagnostic X-ray departments (particularly at its reception area) and other areas where diagnostic X-ray equipment is used, other than dentistry. For example, such a notice might read:

IF IT IS POSSIBLE THAT YOU MIGHT BE PREGNANT, NOTIFY THE PHYSICIAN OR RADIOGRAPHER/TECHNICIAN BEFORE YOUR X-RAY EXAMINATION.

(56) Since fetal doses are usually well below 50 mGy in diagnostic radiology, pregnancy tests are not usually done. In cases where a high-dose fluoroscopy procedure of the abdomen or pelvis (e.g embolisation) is contemplated, depending on the patient reliability and history, the physician may want to order a pregnancy test.

(57) When a patient has been determined to be pregnant or possibly pregnant, a number of steps are usually taken prior to performing the procedure. The technician or clerk who obtains this information should relate this information to the radiologist. The radiologist usually begins by determining whether the conceptus is going to be in the primary X-ray beam. If not, then the risk to the fetus is extremely low and the most important fact is to practice good radiology: i.e., keep the number and type of exposures to a minimum while still getting the correct diagnosis.

(58) If the fetus is going to be in the direct beam, one should ascertain whether the procedure is relatively low-dose (such as a single plain radiograph of the abdomen) or high-dose (fluoroscopy). If the procedure is a high-dose one, it is often helpful to determine whether another type of examination that does not use ionising radiation

(such as ultrasound) can provide the desired diagnostic information. If this is not available, there should be an analysis of the stage of gestation, the estimated fetal dose, the medical indication for the examination, and the risk of delaying the examination. The latter factors depend upon the stage of pregnancy. Often it is helpful to discuss the issues with the referring physician.

(59) Two specific diagnostic examinations require additional discussion, viz. routine chest radiography and X-ray pelvimetry. The World Health Organization, WHO (1992), has concluded that routine (screening) maternal chest radiography during pregnancy is not indicated unless there is a high local incidence of clinically silent chest disease.

(60) Historically, in a number of countries, pelvimetry represented the major single source of ionising radiation to the fetus. While radiographic pelvimetry is sometimes of value, it should be undertaken only on the rare occasions when this is likely to be the case and should not be carried out on a routine basis (WHO, 1999). X-ray pelvimetry provides only limited additional information to physicians involved in the management of labour and delivery. Statistical analysis has indicated a poor correlation between the course of labour and the pelvic measurements. In the few instances in which the clinician thinks that pelvimetry may contribute to a medical treatment decision, the reasons should be clearly delineated.

(61) Presently, there have been attempts to reduce the fetal dose from pelvimetry by using computer techniques (digital radiography) but the equipment is not generally available or used for this purpose. Ultrasonic examinations provide most of the information required by obstetricians and do not utilise ionising radiation.

5.2. During the examination

(62) **When an examination is indicated in which the X-ray beam irradiates the fetus directly, and this cannot be delayed until after pregnancy, care should be taken to minimise the dose to the fetus.**

(63) Medically indicated radiography or fluoroscopy of areas remote from the fetus, such as chest, skull, or extremities (other than the hip), can be done safely at any time during pregnancy if the X-ray equipment is properly shielded and if X-ray beam collimation is used. Tailoring of procedures is usually not necessary.

(64) When pregnant women require abdominal or pelvic diagnostic X-ray examinations in which the X-ray beam irradiates the fetus directly, special care has to be taken to ascertain that the X-ray examination is indeed indicated at that time and that it cannot be delayed until after the pregnancy. Commonly, the radiation risk to the fetus is much less than that of not making a necessary diagnosis. In such cases, care should be taken to minimise the absorbed dose in the fetus. However, alterations in technique should not unduly reduce the diagnostic value of the X-ray examination. (65) Perhaps the most common ways to tailor examinations and reduce fetal exposure are to collimate the beam to a very specific area of interest, increase kVp, remove the anti-scatter grid, or reduce the number of radiographs taken.

(66) A typical example is a pregnant female in whom there is a suspected obstructing distal ureteral stone. Ultrasound may show dilatation of the renal collecting system but it is unlikely to show the level of obstruction or the size of the stone. Rather than performing a routine intravenous urogram (with a preliminary film and then 7 or so sequential post-intravenous contrast films), the diagnosis may often be obtained with one preliminary film and then a single film 10 minutes post contrast administration.

(67) For computed tomography (CT) scans with the uterus in the field of view, the absorbed doses to the fetus are typically about 10–40 mGy. Fortunately, the primary radiation beam on CT scanners is very tightly collimated and can be precisely controlled relative to location by using the scout view. As with other examinations it may be possible to limit the scanning to the anatomical area of interest (for example, the kidneys) rather than to scan the entire abdomen and pelvis.

(68) High fetal doses in diagnostic radiology also arise from abdominal or pelvic fluoroscopy. The radiation dose to the fetus is best controlled by being very careful to minimise the time of exposure. With good technique, fetal doses during a barium enema are in the range of 3–7 mGy. Because of longer fluoroscopy times, doses from double contrast barium enemas are often twice as high as from single contrast studies. If the pregnancy is not recognised, there may not be careful attention to limiting fluoroscopy time. Fetal doses can approach or exceed 50 mGy, especially if the fluoroscopy time exceeds 7 minutes.

(69) When a high-dose procedure is performed and when the fetus is known to be in the primary X-ray beam, the technical factors should be recorded to allow subsequent fetal dose estimation. Important factors are whether a grid was used, the kVp, dose rate, fluoroscopy time, dose-area-product, geometrical description, and projections.

5.3. After irradiation

(70) **For diagnostic radiology, fetal dose estimation is usually not necessary unless the fetus is in the direct beam. Evaluation of fetal doses from pelvic fluoroscopy is subject to more uncertainty than doses from plain radiography or CT.**

(71) Diagnostic irradiation of the pregnant patient can lead to apprehension about possible fetal effects. Even though the absorbed doses to the conceptus are generally small for most diagnostic radiography, such concern may lead to inappropriate suggestions that additional diagnostic examinations be delayed or withheld or even that the pregnancy should be terminated. Because fetal dose from such procedures is almost always less than 100 mGy (the minimum threshold level where radiogenic malformations might occur and the individual probability of radiogenic cancer is very low), fetal irradiation from diagnostic procedures almost never justifies terminating a pregnancy.

(72) After low dose examinations (such as a maternal chest X-ray) in which the conceptus is not in the X-ray beam, there is no real need for individual fetal dose estimations. However, after high-dose abdominal or pelvic CT or fluoroscopy, a

qualified expert should make an estimate of the absorbed dose and the associated risk to the fetus. With such an expert and carefully worded advice, the patient and husband or other appropriate persons should then be in a position to reach their own conclusions. This is discussed further in Section 10. Risk from the radiation exposure will depend primarily upon fetal dose and fetal age.

(73) Determination of the absorbed dose to the embryo or fetus from plain film abdominal or pelvic radiography examinations is difficult, but usually the dose can be estimated within a 50 percent error. For diagnostic radiographic examinations, one can utilise the mean skin exposure per film (for a given examination) and then estimate the absorbed dose at a certain depth if the technical factors concerning the beam energy are known. If the technical factors are not well known, the mean dose to the ovaries often gives an approximation of fetal dose.

(74) Table 1 gives typical uterine or fetal doses for some common routine examinations in the United Kingdom. If the pregnancy was recognised prior to exposure, tailoring of the examination may have reduced these doses.

(75) It should be noted that when dosimetry surveys have been performed within a particular country for diagnostic radiology examinations, doses have been found to vary by a factor of 30 or more for the same examination. This is a function of variations in kVp, waveform, filtration, presence of a grid, film and screen combinations, film processing, and a number of other factors. There is a general tendency to expect that doses from digital fluoroscopy equipment may be lower than with conventional equipment, but in reality this is not often the case. As a result, installation specific measurements and calculations of fetal doses may be necessary if fetal doses are suspected of exceeding 10 mGy.

Table 1. Approximate foetal doses from common diagnostic procedures in the United Kingdom. (Adapted from Sharp, Shrimpton, and Buiy, 1998)

Examination	Mean (mGy)	Maximum (mGy)
Conventional x-ray examinations		
Abdomen	1.4	4.2
Chest	<0.01	<0.01
Intravenous urogram	1.7	10
Lumbar spine	1.7	10
Pelvis	1.1	4
Skull	<0.01	<0.01
Thoracic spine	<0.01	<0.01
Fluoroscopic examinations		
Barium meal (UGI)	1.1	5.8
Barium enema	6.8	24
Computed tomography		
Abdomen	8.0	49
Chest	0.06	0.96
Head	<0.005	<0.005
Lumbar spine	2.4	8.6
Pelvis	25	79

(76) In diagnostic radiology, fetal dose is also significantly affected by patient anatomy including the thickness of the patient, whether the uterus is ante or retroverted, and even the distension of the bladder. While 'fetal dose' is often mentioned and there is the assumption that it is uniform, this is only the case early in pregnancy. As the fetus grows larger, the absorbed dose becomes less uniform. Finally, it is rare for a patient to only have one diagnostic examination and it is useful to see if other examinations were also performed during the gestation.

(77) The estimation of fetal dose after fluoroscopy is more difficult and the range of uncertainty is greater than for routine radiographic examinations. It is extremely difficult to estimate accurately the dose without knowing the fluoroscopy time and the location of the beam. Unfortunately, the fluoroscopy time or other useful parameters are often not recorded. Even if these factors are known, one still cannot be sure how long the conceptus was in the primary beam since the radiologist is usually moving the fluoroscopic beam.

(78) Other factors that affect fluoroscopy dose are whether conventional or pulsed fluoroscopy was used, the magnification mode, and whether a grid was used. Usually these factors are not recorded and can only be estimated based upon the usual practice at that medical facility. In most fluoroscopy cases, a 'best guess' estimate is made and sometimes a 'worst case' estimate of fetal dose. Usually the best estimate and some assessment of the uncertainty are expressed to the patient and the referring physician.

6. NUCLEAR MEDICINE

(79) **Most diagnostic nuclear medicine procedures in developed countries are done with short-lived radionuclides (such as technetium-99^{m}) that do not cause large fetal doses. For radionuclides that do not cross the placenta, fetal dose is derived from the radioactivity in maternal tissues. There are, however, some radiopharmaceuticals (such as iodine isotopes), which do cross the placenta and concentrate in a specific organ or tissue and which can therefore pose significant fetal risks.**

6.1. Before irradiation

(80) **When a nuclear medicine examination is proposed for a pregnant woman, care has to be taken to ascertain that the examination is indeed indicated for a medical condition that requires prompt therapy. For these diagnostic examinations, the risk to the mother of not performing the examination is greater than the radiation risk to the fetus. The possibility of reducing the administered activity should be considered.**

(81) In women of childbearing age, the possibility of pregnancy and the justification for the examination should be considered. The recommended precautions to prevent or minimise irradiation of the fetus include the following:

(82) The patient must be carefully interviewed to assess the likelihood of pregnancy. Particular discretion is required to ascertain the possibility of pregnancy in an adolescent. In order to minimise the frequency of unintentional radiation exposures of the embryo or fetus, advisory notices should be posted at several places within the nuclear medicine department, and particularly at its reception area. For example:

IF IT IS POSSIBLE THAT YOU MIGHT BE PREGNANT, NOTIFY THE PHYSICIAN OR TECHNICIAN BEFORE RECEIVING ANY RADIOACTIVE MATERIAL.

(83) Many patients incorrectly assume that irradiation from a nuclear medicine examination begins when the gamma camera begins imaging, and they may not mention a potential pregnancy until after the radiopharmaceutical has been administered. Therefore, before radiopharmaceutical administration, it is necessary to consider as pregnant any woman of reproductive age presenting for a nuclear medicine examination at a time when a menstrual period is overdue or missed, unless there is information that precludes pregnancy (e.g., hysterectomy or tubal ligation). If the menstrual cycle is irregular, and a non-technetium or therapeutic radiopharmaceutical is being administered, a pregnancy test may be indicated before proceeding.

(84) Many laboratories also ask all females to indicate if they are breast-feeding, since many radiopharmaceuticals can be transferred to a baby via breast milk. Cessation of breast-feeding for at least some period is recommended for most nuclear medicine studies. Breast-feeding is usually stopped for 3 weeks after all ^{131}I and ^{125}I

radiopharmaceuticals except labelled hippurate and after ^{22}Na, ^{67}Ga, and ^{201}Tl. It is stopped for 12 hours after iodine labelled hippurates and all ^{99m}Tc compounds except labelled red blood cells, -phosphonates, and -DTPA, and for at least 4 hours after the latter compounds.

(85). Occasionally, questions arise about the advisability of becoming pregnant after a nuclear medicine examination or treatment. The Commission has recommended that a woman not become pregnant until the potential fetal dose from remaining radionuclides is less than 1 mGy. This is not usually a consideration except for radioiodine therapy or radiopharmaceuticals labelled with ^{59}Fe (for metabolism studies) or ^{75}Se (for adrenal imaging). As a result of the long physical half-lives of these radionuclides and their long residence times in the body, it is recommended that pregnancy be avoided for 6 and 12 months respectively. The special conditions related to radioiodine therapy are discussed later.

6.2. During the diagnostic examination

(86) **Since radionuclides in maternal tissues contribute to fetal dose, maternal hydration and frequent voiding can reduce the fetal dose after the administration of a number of radiopharmaceuticals.**

(87) Irradiation of the fetus results from placental transfer and distribution of radiopharmaceuticals in the fetal tissues, as well as from external irradiation from radioactivity in the mother's organs and tissues. The physical, chemical, and biological properties of the radiopharmaceuticals are the critical factors in possible placental transfer.

(88) Some radiopharmaceuticals cross the placenta freely, e.g., radioactive iodides, and are taken up in fetal tissues, where they irradiate the tissues. Some analogues of natural metabolites (e.g., radiostrontium for calcium and radiocaesium for potassium) are less readily transferred. Radiopharmaceuticals that are retained by the mother, and do not cross the placenta (e.g., radiocolloids), only act as external sources of irradiation to the fetus.

(89) Using smaller administered activities and longer imaging times can reduce the absorbed dose to the fetus. This is feasible if the patient is not too sick and is able to remain still. Occasionally, the sequence of the examinations can be adjusted to reduce radiation dose. A typical example is a ventilation perfusion lung scan ordered on a pregnant patient to exclude a pulmonary embolus. In routine operation, many laboratories will perform the ventilation scan first and then do the perfusion scan. This has advantages in some situations. In the specific case of a suspected pulmonary embolus, the perfusion scan can be performed first, and if it is normal, a ventilation scan is not needed at all.

(90) The choice of radiopharmaceuticals for the ventilation portion of the lung scan can also affect fetal dose. If the scan is performed with ^{133}Xe gas, there is very little fetal dose; however, one can also do ventilation scans using ^{99m}Tc-DTPA aerosol. This will be absorbed and excreted via the kidneys, and while in the bladder it will contribute to fetal dose.

(91) In the case of radiopharmaceuticals that are rapidly eliminated by the maternal kidneys, the urinary bladder, acting as a reservoir, is a major source of fetal irradiation. After the administration of such radiopharmaceuticals, maternal hydration and frequent voiding should, therefore, be encouraged. For those radio-phannaceuticals that have gastrointestinal excretion however, administration of laxatives is only rarely helpful in reducing fetal dose.

(92) Some nuclear medicine patients have pregnant family members at home and enquire about the dose they might give to such a person. Usually the total dose to complete decay from the radionuclide in the patient is calculated at 0.5 or 1.0 meters. For most diagnostic nuclear medicine procedures, the total decay dose at 0.5 meter from the patient ranges from 0.02–0.25 mGy and at 1 meter from the patient the dose is 0.05–0.10 mGy. This poses no significant risk to pregnant family members.

6.3. Nuclear medicine therapy for hyperthyroidism and thyroid carcinoma

(93) **Radioiodine easily crosses the placenta and therapeutic doses can pose significant problems for the fetus, particularly permanent hypothyroidism.**

(94) Because certain radiopharmaceuticals, including ^{131}I as iodide and ^{32}P as phosphate, can rapidly cross the placenta, the possibility of pregnancy should be very carefully considered before such radionuclides are given for therapy or for a whole body ^{131}I scan for thyroid carcinoma. As a rule, a pregnant woman should not be treated with a radioactive substance unless the radionuclide therapy is required to save her life: in that extremely rare event, the potential absorbed dose and risk to the fetus should be estimated and conveyed to the patient and the referring physician. Considerations may include terminating the pregnancy.

(95) In women, thyroid carcinoma comprises over 80% of cancer of the head and neck diagnosed between the ages of 15–45 years. Thyroid cancers are relatively unaggressive compared to most other cancers. As a result both surgical and radio-iodine treatment are often delayed until after pregnancy. In general, if any therapy is to be performed during pregnancy, it will be surgery during the second or third trimester.

(96) Radioiodine will easily cross the placenta and the fetal thyroid begins to accumulate iodine at about 10 weeks of gestational age. Radioiodine therapy is essentially contraindicated in patients who are known to be pregnant. If radioiodine treatment of thyroid carcinoma is to be performed, it should be delayed until after delivery. If this is done, the physician should also be aware that radioiodine is excreted in breast milk and breast-feeding should be stopped completely after a therapeutic dose. If this is not done the infant may become permanently hypothyroid or be at high risk for subsequent thyroid cancer.

(97) A major problem occurs when a female, who is not thought to be pregnant, is treated for thyroid carcinoma and is found out to be pregnant after the administration of radioiodine (cf. paragraphs 107 and 108). Menstrual history is often not adequate to ensure that a patient is not pregnant. In most developed countries, it is common practice to obtain a pregnancy test prior to high-dose ^{131}I scanning or

therapy for women of childbearing age unless there is a clear history of prior tubal ligation or hysterectomy precluding pregnancy. In spite of the above, it still happens that pregnant women are treated, either because of false histories or because the pregnancy is at such an early stage that the pregnancy test is not yet positive.

(98) Most commonly, the pregnancy is early and the major problem is fetal whole body dose due to gamma emissions from radioiodine in the maternal bladder. During pregnancy, the whole body dose to the conceptus is in the range of 50–100 μGy/MBq of administered activity. This dose can be reduced by hydrating the patient and by encouraging frequent voiding.

(99) If the conceptus is more than 8 weeks post conception (and the fetal thyroid may accumulate iodine) and the pregnancy is discovered within 12 hours of iodine administration, giving the mother 60–130 mg of stable potassium iodide (KI) will partially block the fetal thyroid and reduce thyroid dose. After 12 hours post radioiodine administration, this intervention is not very effective.

(100) Maternal hyperthyroidism can occur during pregnancy. The diagnosis can be made on the basis of serum hormone determinations rather than on the basis of radioiodine uptake studies or thyroid scintigraphy. Radioiodine treatment can often be delayed until after pregnancy and the patient treated in the interim with drugs. The major problem again is discovering that a patient is pregnant after they have received a therapeutic dose of radioiodine. The same principles apply that are discussed above.

(101) Most female patients are advised not to become pregnant for at least 6 months after radiotherapy with radioiodine. This is not based upon potential heritable radiation effects, but rather upon the need to be sure (1) that the hyperthyroidism or cancer is controlled, and (2) that another treatment with radioiodine is not going to be needed when the patient is pregnant. It is also based upon the fact the Commission has recommended that enough radioiodine be cleared to ensure that the unborn child not receive a dose in excess of 1 mGy unless it is medically necessary for the health of the mother.

(102) There are occasional circumstances in which ^{32}P, ^{89}Sr, or ^{131}I metaiodobenzylguanidine are used for therapy. In order to keep the dose to the fetus below 1 mGy, pregnancy should be avoided for 3, 24, and 3 months respectively.

(103) Patients treated with radioiodine can be a significant radiation source to pregnant family members. The dose to a family member staying at a distance of 0.5 meters from the patient until the radioactivity totally decays (about 10 weeks) is about 1.3 mGy from a hyperthyroid patient and 6.8 mGy from a thyroid cancer patient. Perhaps more importantly, these patients must also be careful not to transfer radioiodine contamination to family members by direct contact or through indirect means.

6.4. After irradiation

(104) **Careful estimation of fetal doses is not usually necessary after diagnostic nuclear medicine studies involving ^{99m}Tc radiopharmaceuticals. If there has been inadvertent administration of other radiopharmaceuticals (such as radioiodine or gallium), more attention should be given to calculation of the fetal dose and explanation of potential risks.**

(105) As with diagnostic radiology procedures, the pregnant patient can be apprehensive after a procedure has been performed. In the case of nuclear medicine, the patient may be even more apprehensive, realising that an administered radioactive material has been incorporated into her, that it will be there for some time, and that it potentially may cross the placenta to the fetus. As a result of this, more careful explanation to the patient and her husband or other appropriate persons may be needed to put the potential radiation risks into perspective..

(106) In contrast to diagnostic radiography examinations, fetal doses in nuclear medicine depend predominantly upon administered activity, and they are independent of the imaging equipment. Typical uterine/fetal doses for common radiopharmaceuticals are presented in Table 2. The activity for most radiopharmaceuticals is usually measured just prior to administration and recorded. While there may be some individual differences of metabolism and localisation of radiopharmaceuticals in very ill patients, in most cases pregnant women have essentially normal distribution of radiopharmaceuticals and the estimated fetal doses will be reasonably accurate.

(107) If a patient is discovered to be pregnant shortly after a therapeutic radioiodine administration, maternal hydration and frequent voiding should be encouraged to help eliminate maternal radioactivity and to reduce radioiodine residence time in the bladder.

(108) If the pregnancy is discovered within several hours of the radioiodine administration and the fetus is old enough to have a functional thyroid, one should consider giving a thyroid-blocking agent potassium iodide). If the pregnancy is discovered later, the placental transfer of radioiodine can result in very high absorbed doses to the fetal thyroid that may cause significant fetal thyroid damage. Since the fetal whole body dose is usually below 100 mGy, there is no reason to terminate the pregnancy; however, the mother should be given usual levels of replacement thyroid hormone.

Table 2. Fetal whole body dose from common nuclear medicine examinations in early pregnancy and at term. (Dose includes maternal and fetal self dose contributions. Adapted from Russell, Stabin, Sparks et al., 1997, ICRP, 1988, and ICRP, 1998.)

Radiopharmaceutical	Procedure	Administered activity (MBq)	Early (mGy)	9 months (mGy)
^{99m}Tc	Bone scan (phosphate)	750	4.6–4.7	1.8
^{99m}Tc	Lung perfusion (MAA)	200	0.4–0.6	0.8
^{99m}Tc	Lung ventilation (aerosol)	40	0.1–0.3	0.1
^{99m}Tc	Thyroid scan (pertechnetate)	400	3.2–4.4	3.7
^{99m}Tc	Red blood cell	930	3.6–6.0	2.5
^{99m}Tc	Liver colloid	300	0.5–0.6	1.1
^{99m}Tc	Renal DTPA	750	5.9–9.0	3.5
^{67}Ga	Abscess/tumour	190	14–18	25
^{123}I	Thyroid uptake[1)]	30	0.4–0.6	0.3
^{131}I	Thyroid uptake[1)]	0.55	0.03–0.04	0.15
^{131}I	Metastases imaging[1)]	40	2.0–2.9	11.0

[1)] Fetal thyroid doses are much higher than fetal whole body dose, viz. 5–15 mGy/MBq for ^{123}I and 0.5–1.1 Gy/MBq for ^{131}I.

7. RADIOTHERAPY

(109) **In pregnant patients, cancers that are remote from the pelvis usually can be treated with radiotherapy. This however requires careful planning. Cancers in the pelvis cannot be adequately treated by radiotherapy during pregnancy without severe or lethal consequences for the fetus.**

(110) Cancer in pregnancy is relatively uncommon but it constitutes a major problem *for* both physicians and patients. Ionising radiation, chemotherapy, and surgery can all present major risks to the unborn baby. Cancer occurs in about 0.1% of pregnancies and it has been estimated that in the United States alone, about 4,000 women per year are considered for radiotherapy during pregnancy. This number may be decreasing with the development of new effective chemotherapeutic protocols. There are few large studies of the adverse effects of the many treatment regimens.

(111) The ethical and risk/benefit issues for the patient in this setting are quite different from the use of most medical radiation where the patient has both the benefit and the risk. With radiotherapy during pregnancy there is a potential maternal/fetal conflict since the mother would be the major beneficiary while the fetus can be at major risk. The physician also has an ethical balance to achieve in terms of separate responsibilities to the fetus and to the mother.

(112) The most common neoplasms that occur during pregnancy are breast, cervix, leukaemia, lymphoma, melanoma, thyroid, ovary, nasopharynx, oesophagus, and brain cancers. These diverse tumours are managed in quite different ways, and the proximity of the tumour to the fetus is a major determinant in the use of radiotherapy. If the fetus is receiving scattered radiation from the chest, there will be concerns related to potential childhood cancer/leukaemia and, depending on the stage of pregnancy and proximity of the treatment field, perhaps decreased IQ of the child.

(113) For treatment of tumours in the pelvis, the fetus either can be in or very close to the primary beam and effects on the fetus are typically severe (usually fetal death). These potential effects were discussed above in Section 3. Issues related to treatment of thyroid carcinoma with radioiodine were discussed in Sections 6.3 and 6.4.

7.1. Before treatment

(114) **Since fetal doses in radiotherapy can be high, it is important to ascertain whether a female patient is pregnant prior to radiotherapy.**

(115) Pregnancy status may be ascertained on the basis of history, patient age, and prior surgery (such as a hysterectomy or tubal ligation) or through the use of a pregnancy test. Even if a radiotherapy patient is not pregnant, she should be counselled to avoid pregnancy until the potentially harmful radiotherapy or other treatment modalities are concluded and the tumour is cured or adequately controlled.

(116) If a patient is found to be pregnant, there are no hard and fast rules. The decision relative to the treatment course should be an informed one made by the patient, the husband, or other appropriate person(s), the treating oncologist, and other team members (e.g. surgeons, obstetricians, pharmacologists and others such as psychologists). The factors to be considered are many but include at least:

- the stage and aggressiveness of the tumour;
- potential hormonal effects of pregnancy on the tumour;
- various therapies and their length, efficacy, and complications;
- impact of delaying therapy;
- expected effects of maternal ill-health on the fetus;
- the stage of pregnancy;
- fetal assessment and monitoring;
- how and when the baby could be safely delivered;
- whether the pregnancy should be terminated;
- legal, ethical, and moral issues.

(117) While it is difficult to generalise about the adverse effects of chemotherapeutic agents administered during the first trimester of pregnancy, with some drugs up to 10% of exposed fetuses exhibit major malformations. After the first trimester, chemotherapy is not usually associated with teratogenesis or adverse developmental outcome. There is some suggestion that in-utero exposure to chemotherapeutic agents may cause an increase in the risk of pancytopaenia at birth and possibly subsequent neoplasms in the offspring.

(118) The risks of surgery and anesthesia during pregnancy are well known and the major problems are associated with hypotension, hypoxia, and infection. Maternal well-being is also to be considered. Many cancer patients have fevers or infections as a result of the tumour or immunosuppression. There may be an association between hyperthermia and teratogenic effects such as neural tube defects and microphthalmia. The other additional maternal problem that can affect the fetus is malnutrition.

7.2. During radiotherapy

7.2.1. Teletherapy to non-pelvic fields

(119) **Teletherapy to non-pelvic fields during pregnancy can be done, but it requires careful estimation of fetal dose and may require additional shielding.**

(120) A number of cancers occur during pregnancy in locations other than the pelvis or abdomen. Breast cancers complicate about one out of 3,000 pregnancies. This may be treated in a number of ways, including with radiotherapy. Fortunately the radiotherapy is delivered to sites quite distant from the fetus. Usually during radiotherapy, a high-risk obstetrical unit follows such women.

(121) Lymphomas also are relatively common during reproductive years. Literature in the early 1980s suggested the need for a therapeutic abortion if these diseases

presented early in pregnancy. Now, lymphomas can be effectively treated with chemotherapy, and radiotherapy may not be needed at all or may possibly be delayed until late in pregnancy or until after pregnancy.

(122) If radiotherapy is used, it is important to calculate the dose to the fetus before the treatments are given. When external radiotherapy is utilised for treatment of tumours at some distance from the fetus, the most important factor in fetal dose is the distance from the edge of the radiation field. The dose decreases approximately exponentially with distance. Fetal doses for a typical photon treatment regimen for brain cancer are in the range of 30 mGy. For anterior and posterior mantle treatments of the chest for Hodgkin's disease, the dose to portions of an unshielded fetus can be 400–500 mGy.

(123) With ^{60}Co, at distances greater than 10 cm from the field, the dose is higher than with photons because of leakage from the machine head. The dose distribution outside of the primary radiation beam may vary among machines of the same nominal type and energy, as well as with field size. As a result, machine specific measurements should be made.

(124) For ^{60}Co with 10×10 cm fields, Table 3 provides crude estimates of the off-axis dose as a percentage of dose on the central axis. With photons from an accelerator, the percentage of off-axis dose is lower by a factor of about 2–5 (depending upon the photon energy). Usually, dosimetry software programs are very accurate for estimation of tissue dose in the primary treatment field, but uncertainties are much greater at distances outside the field (for example at one meter). In these cases, when dose estimation to peripheral tissues is important, phantom measurements and in-vivo dosimetry are usually used.

(125) Additional shielding can reduce the fetal dose by 50%. However, effective shielding often weighs in the order of 200 kg. It can exceed the design limits for many treatment tables and may cause injury to the patient or technician if not properly constructed and handled.

(126) The American Association of Physicists in Medicine (AAPM) has made a series of recommendations (Stovall, Blackwell, Cundiff et al., 1995), which provide points to be considered:

- Complete all planning as though the patient was not pregnant. If the fetus is near the treatment beam do not take portal localisation films with open collimation and blocks removed.

Table 3. Crude estimates of off-axis dose for ^{60}Co treatments with 10×10 cm fields

Distance from field edge (cm)	Off-axis dose (percent of D_{max} on central axis)
10	1.7
20	0.7
30	0.4
40	0.3
50	0.15

- Consider modifications to the treatment plan that would reduce the radiation dose to the fetus by changing field size, angle, radiation energy, and field trimmers on the edge nearest the fetus. If possible use photon energies of less than 25 MV.
- Estimate dose to the fetus without special shielding, using out of beam phantom measurements at the symphysis pubis, fundus, and a midpoint.
- If fetal dose is above 50–100 mGy, a shield may be constructed with 4–5 half-value layers of lead. Measure dose to fetus in a phantom for simulated treatment with the shielding in place, adjusting radiation amount and location.
- Document the treatment plan and discuss it with the staff involved in patient set-up. Document the shielding (perhaps with a photograph).
- Check weight and load bearing specifications of the treatment couch or other aspects of shielding support.
- Be present during initial treatment to assure that shielding is correctly placed.
- Monitor the fetal size and growth throughout the course of treatment and reassess fetal dose if necessary.
- At completion of treatment, document total dose including range of dose to the fetus during therapy.
- Consider referring patient to another institution if equipment and personnel are not available for reducing and estimating the fetal dose.

7.2.2. Teletherapy and brachytherapy to pelvic fields

(127) **Regardless of protective measures, radiotherapy involving the pelvis of a pregnant female almost always results in severe consequences for the fetus, most likely fetal death.**

(128) Carcinoma of the cervix is the most common malignancy associated with pregnancy. Cervical cancer complicates about one out of 1,250 to 2,200 pregnancies. This rate, however, varies significantly by country. Cervical cancer is often treated by surgery and/or radiotherapy (teletherapy and brachytherapy) and the doses required with both forms of radiotherapy will cause termination of pregnancy. If the tumour is infiltrative and is diagnosed late in pregnancy an alternative is to delay treatment until the baby can be safely delivered.

(129) Ovarian cancer is quite rare during pregnancy, complicating less than one in 10,000 pregnancies. Exploratory surgery is usually employed to make the diagnosis. Most patients with ovarian carcinoma are treated with chemotherapy. Radiotherapy is rarely used to treat this tumour during pregnancy.

(130) A brachytherapy patient is often kept in the hospital until the sources are removed. While such a patient can occasionally be a source of radiation to a pregnant visiting family member, the potential dose to the family member's fetus is very low, irrespective of the type of brachytherapy. Prostate brachytherapy can be done with permanent implantation of radioactive ^{198}Au or ^{125}I 'seeds', and the patient is discharged from the hospital with these in place. The short range of the emissions from these radionuclides is the reason that the patient can be discharged and is the reason that these patients pose no danger to pregnant family members.

7.3. After radiotherapy

(131) **After radiotherapy involving a pregnant patient, careful records of the technique and fetal dose estimation should be maintained. Since there may be fetal consequences, careful counselling and follow-up is recommended.**

(132) Since radiotherapy usually involves treatment over several weeks, pregnancy is usually identified before or during treatment. It is extremely rare for a patient to receive a full treatment course of teletherapy or brachytherapy and then be discovered to be pregnant afterwards. Even with prior counselling and appropriate shielding during treatment, the patient will often want additional information.

(133) The final estimates of the fetal dose should be calculated and documented. This should include details about the technical factors discussed above. An appropriately trained medical physicist should do such calculations and the potential risks should be conveyed to the mother. Although local regulations vary, it is often necessary to keep these records for many years and usually until the child becomes an adult.

(134) Occasionally, patients who are not pregnant ask when they can become pregnant after radiotherapy. Most radiation oncologists request that their patients not become pregnant for 1–2 years after completion of therapy. This is not primarily related to concerns about potential radiation effects, but rather to considerations about the risk of relapse of the tumour that would require more radiation, surgery, or chemotherapy.

8. MANAGEMENT OF PREGNANT PHYSICIANS AND OTHER STAFF

(135) **The basis for the control of the occupational exposure of women who are not pregnant is the same as that for men. However, if a woman is, or may be, pregnant, additional controls have to be considered to protect the unborn child. There is, therefore, a recommended fetal dose limit. There are a number of ways in which compliance with this limit may be achieved.**

(136) Dose limits for the fetus are broadly comparable with those for the general public. This is reasonable since while the mother may have chosen to be a radiation worker, the unborn child has not made such a decision. The adoption of a rigid dose limit for the conceptus of a pregnant woman who is occupationally exposed would pose practical problems. The early part of pregnancy (before the pregnancy has been declared) is covered by the normal protection of workers, which is essentially the same for males and females.

(137) Once the pregnancy has been declared, and the employer notified, additional protection of the fetus should be considered. The working conditions of a pregnant worker, after the declaration of pregnancy, should be as such to make it unlikely that the additional dose to the conceptus will exceed about 1 mGy during the remainder of pregnancy. In the interpretation of this recommendation, it is important not to create unnecessary discrimination against pregnant women. There are responsibilities on both the worker and the employer. The first responsibility for the protection of the conceptus lies with the woman herself to declare her pregnancy to the management as soon as the pregnancy is confirmed.

(138) The restriction on dose to the conceptus does not mean that it is necessary for pregnant women to avoid work with radiation or radioactive materials completely, or that they must be prevented from entering or working in designated radiation areas. It does, however, imply that their employer should carefully review the exposure conditions of pregnant women. In particular, their employment should be of such a type that the probability of high accidental doses and radionuclide intakes is insignificant.

(139) When a medical radiation worker is known to be pregnant, there are three options that are often considered in medical radiation facilities: 1) no change in assigned working duties; 2) change to another area where the radiation exposure may be lower; or 3) change to a job that has essentially no radiation exposure. There is no one correct answer for all situations, and in certain countries there may be specific regulations. It is desirable to have a discussion with the employee. The worker should be informed of the potential risks, local policies, and recommended dose limits.

(140) Change to a position where there is no radiation exposure is sometimes requested by pregnant workers who realise that risks may be small but do not wish to accept any increased risk. The employer may also arrange for this in order to avoid future difficulties in case the employee delivers a child with a spontaneous congenital abnormality (which occurs at a rate of about 3 in every 100 births). This approach is not required on a radiation protection basis, and it obviously depends

on the facility being sufficiently large and flexible to have other employees fill the vacated position.

(141) Change to a position that may have lower ambient exposure is also a possibility. In diagnostic radiology, this may involve transfer of a technician from fluoroscopy to CT scanning or some other area where there is less scattered radiation to workers. In nuclear medicine departments, a pregnant technician can be restricted from spending a lot of time in the radiopharmacy or working with solutions of radioiodine. In radiotherapy with sealed sources, pregnant technicians or nurses might not participate in manual brachytherapy.

(142) An ethical consideration is involved in both of these last two alternatives since another worker will have to incur additional radiation exposure because a co-worker became pregnant.

(143) There are many situations in which the worker wishes to continue doing the same job, or the employer may depend on the worker to continue in order to provide adequate patient care. From a radiation protection point of view, this is perfectly acceptable providing the fetal dose can be reasonably accurately estimated and falls within the recommended limit of 1 mGy fetal dose after the pregnancy is declared. It would be reasonable to evaluate the work environment in order to provide assurance that high-dose accidents are unlikely.

(144) The recommended dose limit applies to the fetal dose and it is not directly comparable to the dose measured on a personal dosimeter. A personal dosimeter worn by diagnostic radiology workers may overestimate fetal dose by about a factor of 10 or more. If the dosimeter has been worn outside a lead apron, the measured dose is likely to be about 100 times higher than the fetal dose. Workers in nuclear medicine and radiation therapy usually do not wear lead aprons and are exposed to higher photon energies. In spite of this, fetal doses are not likely to exceed 25 percent of the personal dosimeter measurement.

(145) Finally, factors other than radiation exposure should be considered in evaluating pregnant workers' activities. In a medical setting there are often requirements for lifting patients and for stooping or bending below knee level. There are a number of national groups that have established non-radiation related guidelines for such activities at various stages of pregnancy.

(146) Occasionally, there are situations where family members provide essential medical care, either in the hospital or at home, to patients who have received radionuclides. In such circumstances, public dose limits do not apply to such a family member. Efforts should optimally be directed at not involving females who are or might potentially be pregnant. If it is essential to involve the help of a pregnant female, it should be done in such a way that the fetal dose from this involvement will not exceed 1 mGy.

9. PREGNANCY AND BIOMEDICAL RESEARCH INVOLVING RADIATION EXPOSURE

(147) **In many countries, radiation exposure of pregnant females in biomedical research is not specifically prohibited. However, their involvement in such research is very rare and should be discouraged.**

(148) The Commission has published the general principles of radiological protection in biomedical research in ICRP *Publication 62* (1991b). The potential harm to the embryo and fetus from medical irradiation was outlined above in the relevant sections on diagnostic radiology, nuclear medicine, and radiotherapy.

(149) Pregnant women should not be involved in biomedical research projects involving radiation exposure unless the pregnancy itself is central to the research and only if alternative techniques involving less risk cannot be used. Even in such a situation, there remains a very difficult ethical issue if a pregnant female receives radiation exposure while serving as a control subject in a research project.

(150) All human research should follow the recommendations of the World Medical Assembly published in the Declaration of Helsinki of 1964 (updated in Tokyo 1975), which is reproduced in ICRP *Publication 62*. All subjects should be volunteers and should have given appropriate informed consent. The potential risks should be predictable and assessed.

10. CONSIDERATION OF TERMINATION OF PREGNANCY AFTER RADIATION EXPOSURE

(151) **Termination of pregnancy is an individual decision affected by many factors. Fetal doses below 100 mGy should not be considered a reason for terminating a pregnancy. At fetal doses above this level, there can be fetal damage, the magnitude and type of which is a function of dose and stage of pregnancy.**

(152) Medical staff may attempt to identify pregnant patients, in order to avoid unnecessary radiation exposure and reduce necessary exposure; yet there are pregnant patients who need to be exposed to radiation. Employers may arrange to keep occupationally exposed pregnant women from exceeding recommended limits; even so, these pregnant workers will still be exposed to some radiation. In any of these circumstances, the pregnant female may be extremely concerned about the outcome of the pregnancy, and a counselling session with the mother and father is often useful.

(153) The issue of pregnancy termination is undoubtedly managed differently around the world. It is complicated by individual ethical, moral, and religious beliefs as well as perhaps being subject to laws or regulations at a local or national level. This complicated issue involves much more than radiation protection issues. This document is intended to provide information that can be helpful in counselling the patient and partner.

(154) Counselling can be accomplished after attempting to estimate the dose to the conceptus from the procedure and comparing radiation risk with the other risks of pregnancy. Women exposed even to low levels of ionising radiation often imagine that they have a much higher risk of malformations than the naturally occurring risk, but appropriate counselling can be beneficial. One useful approach is to indicate to the patient the probability of *not* having a child with either a malformation or cancer, and how that probability is affected by radiation (Table 4).

(155) With the exception of radiotherapy of the abdomen or pelvis, and major accidents, the magnitude of effects that may occur from medical radiation is generally small compared with the normal incidence of other problems during pregnancy. In a non-exposed population (i.e., exposed only to natural background radiation), approximate risks during pregnancy include a 15 percent or greater spontaneous abortion rate, a 2 to 4 percent major malformation incidence, a 4 percent intrauterine growth retardation rate (mostly due to hypertension), and an 8 to 10 percent incidence of genetic diseases.

(156) For fetal doses less than 100 mGy, there is no medical justification for terminating a pregnancy because of radiation exposure. A conservative estimate of the lifetime risk of radiogenic induction of childhood cancer or leukaemia at 100 mGy is about 1 in 170. Without radiation exposure (apart from natural background) the lifetime risk of contracting cancer is about 1 in 3; for fatal cancer the risk is about 1 in 5. As pointed out earlier, malformations due to radiation probably do not occur at fetal doses less than 100–200 mGy.

(157) When the dose to the fetus exceeds 100–200 mGy, the approach to the problem is slightly different. This situation may involve radiotherapy or accidental

Table 4. Probability of bearing healthy children as a function of radiation dose

Absorbed dose to conceptus, mGy, above natural background	Probability that child will have ***no*** malformation, %	Probability that child will ***not*** develop cancer (age 0-19), %[1)]
0	97	99.7
0.5	97	99.7
1.0	97	99.7
2.5	97	99.7
5	97	99.7
10	97	99.6
50	97	99.4
100	(close to 97)[2)]	99.1

[1)] Rounded values. Radiation risk for fatal cancer conservatively assumed to be 0.6% per 100 mGy fetal dose, corresponding to about 1/17,000 per mGy, and a linear dose–response relationship. Many epidemiological studies suggest that the risk may be lower than that assumed here. Background risk of childhood cancer calculated from NCI-SEER (1994).

[2)] Although the exact risk in humans is uncertain, animal data suggest that malformations due to radiation are not likely at doses less than 100–200 mGy. Above this malformations would only be observed if exposure were between the 3rd and 25th weeks of gestation. The risk of malformation is low at 100–200 mGy but will increase with increasing dose. Decreased IQ and possible retardation are only detectable when foetal doses exceed 100 mGy during the 8th to 25th weeks of gestation.

rather than diagnostic exposure, and the estimates of absorbed fetal dose may have a larger factor of uncertainty. If the fetal absorbed dose is high, e.g., in excess of 500 mGy, and it was incurred during the 3rd to 16th weeks of conception, there is a substantial chance of growth retardation and central nervous system damage. Although it is possible that the fetus may survive doses in this range, the parents should be informed of the high risks involved.

(158) In the intermediate dose range, 100–500 mGy, the situation is less clear-cut, although such circumstances arise relatively infrequently. In this absorbed dose range the risk of a measurable reduction in IQ must be seriously considered if the fetus was exposed between 8 and 15 weeks of gestational age. In such instances, a qualified biomedical or health physicist should calculate the absorbed fetal dose as closely as possible, and the physician should ascertain the individual and personal situation of the parents. For example, if the dose to the fetus was estimated to be just above 100 mGy and the parents had been trying to have a child for several years, they may not wish to terminate the pregnancy. This should be a personal decision made by the parents after being appropriately informed.

11. SUMMARY RECOMMENDATIONS

(159) Medical professionals using radiation should be familiar with the effects of radiation on the embryo and fetus. At most diagnostic levels this would include risk of childhood cancer, while at doses in excess of 100–200 mGy risks related to nervous system abnormalities, malformations, growth retardation, and fetal death should be considered. The magnitude of these latter risks differs quite considerably between the various stages of pregnancy.

(160) All medical practices (both occupational and patient-related) involving radiation exposure should be justified (result in more benefit than risk). Medical exposures should also be justified on an individual basis. This includes considerations that balance medical needs against potential radiation risks. This is done using judgement rather than numerical calculations. Medical exposure of pregnant women poses a different benefit/risk situation than most other medical exposures. In most medical exposures the benefit and risk are to the same individual. In the situation of in-utero medical exposure there are two different entities (the mother and the fetus) who must be considered.

(161) Prior to radiation exposure, female patients in the childbearing age group should be evaluated and an attempt made to determine who is or could be pregnant.

(162) Medical radiation applications should be optimised to achieve the clinical purposes with no more radiation than is necessary, given the available resources and technology. If possible, for pregnant patients, the medical procedures should be tailored to reduce fetal dose.

(163) After medical procedures involving high doses of radiation have been performed on pregnant patients, fetal dose and potential fetal risk should be estimated.

(164) Pregnant medical radiation workers may work in a radiation environment as long as there is reasonable assurance that the fetal dose can be kept below 1 mGy during the course of pregnancy.

(165) Radiation research involving pregnant patients should be discouraged.

(166) Termination of pregnancy at fetal doses of less than 100 mGy is not justified based upon radiation risk. At higher fetal doses, informed decisions should be made based upon individual circumstances.

BIBLIOGRAPHY

Adams, E., Brettle, D., Jones, A. et al. (1997) Estimation of foetal and effective dose for CT examinations. Brit. J. Radiol. 70, 272–278.

Allen, H., Nisker, J. (Eds.) (1986) *Cancer in Pregnancy: Therapeutic Guidelines.* Futura Publishing, Mt. Kisco, NY, USA.

American Medical Association, Council on Scientific Affairs (1984) Effects of pregnancy on work performance. J. American Medical Association 251, 1995–1997.

Boice, J.D. Jr., Miller, R.W. (1999) Childhood and adult cancer after intrauterine exposure to ionising radiation. Teratology 59, 227–233.

Brent, R.L. (1986) Radiation teratogenesis. In: Sever, J.L., Brent, R.L. (Eds.), *Teratogen Update: Environmentally Induced Birth Defect Risks.* A.R. Liss, New York, pp. 145–163.

Brent, R.L., Meistrich, M., Paul, M. (1993) Ionizing and non-ionizing radiations. In: Paul, M. (Ed.), *Occupational and Envireonmental Reproductive Hazards.* Williams and Wilkins, Baltimore, MD, USA, pp. 165–189.

Brent, R.L. (1994) The effect of embryonic and fetal exposure to x-ray, microwaves, ultrasound, magnetic resonance, and isotopes. In: Barron, W.M., Lindheimer, M.D. (Eds.), *Medical Disorders during Pregnancy*, 2nd ed. Mosby Yearbook, Inc., St. Louis, MO, USA, pp. 487–518.

Brent, R.L. (1992) The effects of embryonic and fetal exposure to ionizing radiation: Counseling the patient and worker about these risks. In: Mossman, K.L., Mills, W.A. (Eds.), *The Biological Basis of Radiation Protection Practice.* Williams and Wilkins, Baltimore, MD, USA, pp. 23–62.

Brent, R.L. (1995) Ionizing Radiation. In: Queenan, J.T., Hobbins, J.C. (Eds.), *Protocols for High Risk Pregnancies,* 3rd ed. Blackwell Scientific, Cambridge, MA pp 26–36.

Brent, R.L., Beckman, D.A. (1996) Developmental effects following radiation of embryonic and fetal exposure to x-ray and isotopes: Counselling the pregnant and nonpregnant patient about these risks. In: Hendee, W.K., Edwards, F.M. (Eds.), *Health Effects of Low Level Exposure to Ionizing Radiation.* Institute of Physics Publishing, Bristol and Philadelphia, pp. 169–2 13.

Damilakis, J., Perisinakis, K., Granimatikakis, J. et al. (1996) Accidental embryo irradiation during barium enema examinations: An estimation of absorbed dose. Investigative Radiology 31, 242–245.

Delongchamp, R.R., Mabuchi, K., Yoshimoto, Y. et al. (1997) Cancer mortality among atomic bomb survivors exposed in utero or as young children, October 1950-May 1992. Radiat Res. 147, 385–395.

Doll, R., Wakeford, R. (1997) Risk of childhood cancer from foetal irradiation. Brit. J. Radiol. 70, 130–139.

European Commission, DG XI (1999) *Guidance for Protection of Unborn Children and Infants Irradiated due to Parental Medical Exposures.* Radiation Protection, No. CR-18-98-316-EN-C. Official Publications of the European Communities, Luxembourg.

Goldman, SM., Wagner, L.K. (1996) Radiologic management of abdominal trauma in pregnancy. Am. J. Roentgenol. 166, 763–767.

ICRP (1986) Developmental effects of irradiation on the brain of the embryo and fetus. ICRP Publication 49, Annals of the ICRP 16 (4).

ICRP (1988) Radiation dose to patients from radiopharmaceuticals. ICRP Publication 53, Annals of the ICRP 18 (1–4).

ICRP (1991a) 1990 Recommendations of the International Commission on Radiological Protection. ICRP Publication 60, Annals of the ICRP 21(1–3).

ICRP (199lb) Radiological protection in biomedical research. ICRP Publication 62, Annals of the ICRP 22 (3).

ICRP (l991c) Summary of the current ICRP principles for protection of the patient in diagnostic radiology. In: ICRP Publication 62, Annals of the ICRP 22 (3).

ICRP (1994) Summary of the current ICRP principles for protection of the patient in nuclear medicine. In: ICRP Publication 68, Annals of the ICRP 24 (4).

ICRP (1996) Radiological protection and safety in medicine. ICRP Publication 73, Annals of the ICRP 26 (2).

ICRP (1997) General principles for the radiation protection of workers. ICRP Publication 75, Annals of the ICRP 27 (1).

ICRP (1998) Radiation dose to patients from radiopharmaceuticals. Addendum to ICRP 53. ICRP Publication 80, Annals of the ICRP 28 (3).

Jensh, R.P., Eisenman, L.M., Brent, R.L. (1995) Postnatal neurophysiologic effects of prenatal x-irradiation. Int. J. Radial. Biol. 67, 217–227.

Konermann, G. (1987) Postimplantation defects in development following ionizing irradiation. In: Nygaard, O.F., Sinclair, W.K., Lett, J.T. (Eds.), *Advances in Radiation Biology, Vol. 13.* Academic Press, San Diego, New York, Boston, London, Sydney, Tokyo, Toronto, pp. 91–167.

Koren, G., Lishner, M., Fame, D. (Eds.) (1996) *Cancer in Pregnancy: Maternal and Foetal Risks.* Cambridge University Press, Cambridge, UK.

Lavoie, C., Don, C. (1997) In-vivo measurement method of ovarian dose during barium enema examinations. Brit. J. Radiol 70, 291–295.

Miller, R.W., Boice, J.D. Jr. (1997) Cancer after intrauterine exposure to the atomic bomb. Radiat. Res. 147, 396–397.

NCI-SEER (1994) *Surveillance, Epidemiology and End Results Cancer Statistics Review 1973–1991: Tables and Graphs.* National Cancer Institute, Bethesda, MD, USA.

NCRP (1977) *Medical Exposure of Pregnant and Potentially Pregnant Women. NCRP Report No. 54.* National Council on Radiation Protection and Measurements, Bethesda, MD, USA.

NCRP (1996) *Sources and Magnitude of Occupational and Public Exposures from Nuclear Medicine Procedures. NCRP Report No. 124.* National Council on Radiation Protection and Measurements, Bethesda, MD, USA.

NCRP (1998) *Radionuclide Exposure of the Embryo/Fetus, NCRP Report No. 128.* National Council on Radiation Protection and Measurements, Bethesda, MD, USA.

Russell, J.R., Stabin, M.G., Sparks, R.B. (1997) Placental transfer of radiopharmaceuticals and dosimetry in pregnancy. Health Phys. 73, 747–755.

Russell, J.R., Stabin, MG., Sparks, R.B. et at. (1997) Radiation absorbed dose to the embryo/fetus from radiopharmaceuticals. Health Phys. 73, 756–769.

Sharp, C., Shrimpton, J.A., Bury, R.F. (1998) *Diagnostic Medical Exposures: Advice on Exposure to Jonising Radiation during Pregnancy*. National Radiological Protection Board, Chilton, Didcot, Oxon, UK.

Sinclair, W.K., Adeistein, S.J., Brent, R.L. et al. (1994) *Considerations Regarding the Unintended Exposure of the Embryo, Fetus, or Nursing Child. NCRP Commentary No. 9*. National Council on Radiation Protection and Measurements, Bethesda MD, USA.

Sneed, P., Albright, N., Wara, W. et al. (1995) Fetal dose estimates for radiotherapy of brain tumours during pregnancy. Int. J. Radiat. Oncol. Biol. and Physics 32, 823–830.

Stovall, M., Blackwell, C., Cundiff, I. et al. (1995) Fetal dose from radiotherapy with photon beams: Report of the AAPM Radiation Therapy Task Group No. 36. Med. Phys. 22, 63–82.

Strahlenschutzkommission (1989) *Effects of Prenatal Irradiation*. Veroffentlichungen der Strahlenschutzkommission, Band 2. Gustav Fischer-Verlag, Stuttgart, New York.

Streffer, C., Molls, M. (1987) Cultures of preimplantation mouse embryos: A model for radiobiological studies. In: Nygaard, OF., Sinclair, W.K., Lett, J.T. (Eds.), *Advances in Radiation Biology, Vol. 13*. Academic Press, San Diego, New York, Boston, London, Sydney, Tokyo, Toronto, pp. 169–213.

UNSCEAR (1993) *Sources and Effects of Ionizing Radiation. United Nations Scientific Committee on the Effects of Atomic Radiation 1993 Report to the General Assembly, Annex H: Radiation effects on the developing human brain*. UN Publication Sales No. E.94.IX.2. United Nations, New York, NY, USA.

Wagner, L., Lester, R., Saldana, L. (Eds.) (1997) *Exposure of the Pregnant Patient to Diagnostic Radiations: A Guide to Medical Management* (2nd ed). Medical Physics Publishing, Madison, WI, USA.

WHO (1992) *Antenatal Care and Maternal Health: How Effective is it? A Review of Evidence*. World Health Organization MSM 92.4.

WHO (1999) *Care in Normal Birth: A Practical Guide*. World Health Organization. FRH/MSM/96-24, on-line version 99.

Yoshimoto, Y., Delongchamp, R., Mabuchi, K. (1994) In utero exposed atomic bomb survivors: cancer risk update. Lancet 344, 345–346.

International Commission on Radiological Protection:

History, policies, procedures

A free booklet describing ICRP

The International Commission on Radiological Protection (ICRP) recently issued this document as a straightforward review of its history, mode of operations, concepts, and current policies. After an Introduction, it presents the History and affiliation and The structure and mode of operation of the Commission. The objectives of the Commission's recommendations are laid out; Quantities, The biological basis of the Commission's policy, Concepts, and The quantitative basis for risk estimates are briefly explained. The structure of the system of protection, Problems of interpretation and application in the system of protection, and The need for stability, consistency, and clarity are all briefly discussed.

This booklet is not part of the *Annals of the ICRP*, and is not automatically distributed to subscribers.

Individual copies may be requested free of charge from

The Scientific Secretary, Dr J Valentin
ICRP
SE-171 16 Stockholm
Sweden.

Bulk deliveries (5 or more copies) can be arranged through the ICRP Scientific Secretary, Dr Valentin, at £ 1.00 per copy.

The International Commission on Radiological Protection

Our Mission Statement

The International Commission on Radiological Protection, ICRP, is an independent Registered Charity, established to advance for the public benefit the science of radiological protection, in particular by providing recommendations and guidance on all aspects of protection against ionising radiation.

The primary body in radiological protection is ICRP. It was formed in 1928 as the 'International X-ray and Radium Committee', but adopted its present name in 1950 to reflect its growing involvement in areas outside that of occupational exposure in medicine, where it originated.

ICRP consists of the Main Commission, Committee 1 (Radiation Effects), Committee 2 (Doses from Radiation Exposure), Committee 3 (Protection in Medicine), Committee 4 (Application of ICRP Recommendations), *ad hoc* Task Groups and Working Parties, and the Scientific Secretariat.

The Main Commission consists of 12 members and a Chairman, while the Committees contain between 15 and 20 members each. The Commission and its Committees run for four-year periods, from 1 July. On each occasion of a new period, at least three, and not more than five, members of the Commission must be changed. A similar rate of renewal is sought for the Committees. Such a new period began 1 July 1997.

The Commission meets once or twice a year. Each Committee meets once a year. Twice in each four-year period, the annual meeting of the Committees is conducted jointly and together with the Commission. These meetings are funded as necessary from monies available to ICRP.

The activities of ICRP are financed mainly by voluntary contributions from national and international bodies with an interest in radiological protection. Some additional funds accrue from royalties on ICRP *Publications*. Members' institutions also provide support to ICRP by making the members' time available without charge and, in many cases, contributing to their costs of attending meetings.

The Commission uses Task Groups and Working Parties to deal with specific areas. Task Groups are formally appointed by the Commission to perform a defined task, usually the preparation of a draft report. A Task Group usually contains a majority of specialists from outside the Commission's structure. It is funded as necessary from monies available to ICRP.

Working Parties are set up by Committees to develop ideas, sometimes leading to the establishment of a Task Group. The membership of a Working Party is usually limited to Committee members. Working Parties receive no funding of their own, *i.e.* they operate primarily by correspondence and by meetings in direct conjunction with meetings of the Committee concerned.

These activities are co-ordinated with a minimum of bureaucracy by a Scientific Secretary, ensuring that ICRP recommendations are promulgated.

Thus, ICRP is an independent international network of specialists in various fields of radiological protection. At any one time, about 100 eminent scientists are actively involved in the work of ICRP. The four-tier structure described provides a rigorous Quality Management system of peer review for the production of ICRP Publications.

In preparing its recommendations, the Commission considers the fundamental principles and quantitative bases on which appropriate radiation protection measures can be established, while leaving to the various national protection bodies the responsibility of formulating the specific advice, codes of practice, or regulations that are best suited to the needs of their individual countries. The aim of the recommendations of ICRP is to

—*provide an appropriate standard of protection for mankind from sources of ionising radiation, without unduly limiting beneficial practices that give rise to exposure to radiation.*

Composition of the International Commission on Radiological Protection and Committees, 1997–2001

COMMITTEE 3
Chairman:
Professor F Mettler

Members:
Professor J M Cosset
Dr M J Guiberteau
Dr L K Harding
Professor Dr J Liniecki
Professor S Mattsson
Professor Nakamura H
Dr P Ortiz
Dr L V Pinillos-Ashton
Professor M M Rehani
Professor H Ringertz
Dr M Rosenstein
Dr Sasaki Y
Dr C Sharp
Professor Yin W
Professor W Y Ussov

COMMITTEE 4
Chairman:
Mr B C Winkler

Members:
Dr R M Alexakhin
Dr E D'Amato
Dr D Cancio
Mr T Godås
Dr A J González
Dr K R Kase
Dr Kosako T
Professor Dr W Kraus
Mr J Lochard
Dr A C McEwan
Dr R V Osborne
Dr K C Pillai
Dr A Sugier
Dr J E Till
Dr A D Wrixon
Dr Xia Y
Dr C Zuur

PAST PUBLICATIONS OF THE *ANNALS OF THE ICRP*

ublications of the ICRP
ull details of all ICRP reports can be obtained from your nearest Elsevier Science office.

CRP CD-ROMS
he ICRP Database of Dose Coefficients: Workers and Members of the
ublic (Version One, 1999) 0 08 042751 0

ublished reports of the ICRP
CRP Publication 83 (Annals of the ICRP Vol. 29 No. 3–4, 1999)
isk Estimation for Multifactorial Diseases 0 08 043900 4
CRP Publication 82 (Annals of the ICRP Vol. 29 No. 1–2, 1999)
rotection of the Public in Situations of Prolonged Radiation Exposure 0 08 043898 9
CRP Publication 81 (Annals of the ICRP Vol. 28 No. 4, 1998)
adiation Protection Recommendations as Applied to the Disposal of
ong-Lived Solid Radioactive Waste 0 08 043859 8
CRP Publication 80 (Annals of the ICRP Vol. 28 No. 3, 1998)
adiation Dose to Patients from Radiopharmaceuticals 0 08 043573 4
CRP Publication 79 (Annals of the ICRP Vol. 28 No. 1–2, 1998)
enetic Susceptibility to Cancer 0 08 042752 9
CRP Publication 78 (Annals of the ICRP Vol. 27 No. 3–4, 1998)
dividual Monitoring for Internal Exposure of Workers: Replacement of ICRP
ublication 54 0 08 042750 2
he ICRP Database of Dose Coefficients: Workers and Members of the Public
CD-ROM) 0 08 004278 0
CRP Publication 77 (Annals of the ICRP Vol. 27 Supplement, 1998)
adiological Protection Policy for the Disposal of Radioactive Waste 0 08 042749 9
CRP Publication 76 (Annals of the ICRP Vol. 27 No. 2, 1997)
rotection from Potential Exposures: Application to Selected Radiation Sources 0 08 0427448 8
CRP Publication 75 (Annals of the ICRP Vol. 27 No. 1, 1997)
eneral Principles for the Radiation Protection of Workers 0 08 042741 3
CRP Publication 74 (Annals of the ICRP Vol. 26 No. 3–4,1996)
*onversion Coefficients for use in Radiological Protection against External Radiation*0 08 042739 1
CRP Publication 73 (Annals of the ICRP Vol. 26 No. 2, 1996)
adiological Protection and Safety in Medicine 0 08 042738 3
CRP Publication 72 (Annals of the ICRP Vol. 26 No. 1, 1996)
ge-dependent Doses to Members of the Public from Intake of Radionuclides: Part 5
ompilation of Ingestion and Inhalation Dose Coefficients 0 08 042737 5
CRP Publication 71 (Annals of the ICRP Vol. 25 No. 3–4, 1996)
ge-dependent Doses to Members of the Public from Intake of Radionuclides: Part 4
halation Dose Coefficients 0 08 042736 7
CRP Publication 70 (Annals of the ICRP Vol. 25 No. 2, 1995)
asic Anatomical and Physiological Data for use in Radiological Protection:
he Skeleton 0 08 042665 4
CRP Publication 69 (Annals of the ICRP Vol. 25 No. 1, 1995)
ge-dependent Doses to Members of the Public from Intake of Radionuclides: Part 3
gestion dose coefficients 08 042658 1

ICRP Publication 68 (Annals of the ICRP Vol. 24 No. 4, 1995)
Dose Coefficients for Intakes of Radionuclides by Workers 0 08 042651
ICRP Publication 67 (Annals of the ICRP Vol. 23 No. 3/4, 1994)
Age-dependent Doses to Members of the Public from Intake of Radionuclides: Part 2 Ingestion dose coefficients 0 08 041155
ICRP Publication 66 (Annals of the ICRP Vol. 24 No. 1–3, 1994)
Human Respiratory Tract Model for Radiological Protection 0 08 041154
ICRP Publication 65 (Annals of the ICRP Vol. 23 No. 2, 1994)
Protection Against Radon-222 at Home and at Work 0 08 042475
ICRP Publication 64 (Annals of the ICRP Vol. 23 No. 1, 1993)
Protection from Potential Exposure: A Conceptual Framework 0 08 042205
ICRP Publication 63 (Annals of the ICRP Vol. 22 No. 4, 1993)
Principles for Intervention for Protection of the Public in a Radiological Emergency 0 08 042204
ICRP Publication 62 (Annals of the ICRP Vol. 22 No. 3, 1993)
Radiological Protection in Biomedical Research 0 08 042203
Annals of the ICRP Vol. 22 No. 1, 1992
Risks Associated with Ionising Radiations 0 08 041840
ICRP Publication 61 (Annals of the ICRP Vol. 21 No. 4, 1991)
Annual Limits on Intake of Radionuclides by Workers Based on the 1990 Recommendations 0 08 041145
ICRP Publication 60 (Annals of the ICRP Vol. 21 No. 1–3,1992)
1990 Recommendations of the International Commission on Radiological Protection 0 08 041144
ICRP Publication 59 (Annals of the ICRP Vol. 22 No. 2, 1992)
The Biological Basis for Dose Limitation in the Skin 0 08 041143
ICRP Publication 58 (Annals of the ICRP Vol. 20 No. 4, 1990)
RBE for Deterministic Effects 0 08 040173
ICRP Publication 57 (Annals of the ICRP Vol. 20 No. 3, 1990)
Radiological Protection of the Worker in Medicine and Dentistry 0 08 040769
ICRP Publication 56 (Annals of the ICRP Vol. 20 No. 2, 1990)
Age-dependent Doses of the Public from Intake of Radionuclides: Part 1 0 08 040763
ICRP Publication 55 (Annals of the ICRP Vol. 20 No. 1, 1989)
Optimization and Decision-making in Radiological Protection 0 08 037388
ICRP Publication 54 (Annals of the ICRP Vol. 19 No. 1–3, 1989)
Individual Monitoring for Intakes of Radionuclides by Workers: Design and Interpretation 0 08 035600
ICRP Publication 53 (Annals of the ICRP Vol. 18 No. 1–4, 1988)
Radiation Dose to Patients from Radiopharmaceuticals 0 08 035591
ICRP Publication 52 (Annals of the ICRP Vol. 17 No. 4, 1988)
Protection of the Patient in Nuclear Medicine 0 08 033188
ICRP Publication 51 (Annals of the ICRP Vol. 17 No. 2/3, 1988)
Data for Use in Protection Against External Radiation 0 08 035587
ICRP Publication 50 (Annals of the ICRP Vol. 17 No. 1, 1987)
Lung Cancer Risk from Indoor Exposures to Radon Daughters 0 08 035579
ICRP Publication 49 (Annals of the ICRP Vol. 16 No. 4, 1987)
Developmental Effects of Irradiation on the Brain of the Embryo and Fetus 0 08 035203
ICRP Publication 48 (Annals of the ICRP Vol. 16 No. 2/3, 1986)
The Metabolism of Plutonium and Related Elements 0 08 034827
ICRP Publication 47 (Annals of the ICRP Vol. 16 No. 1, 1986)
Radiation Protection of Workers in Mines 0 08 034020

:RP Publication 46 (Annals of the ICRP Vol. 15 No. 4, 1986)
adiation Protection Principles for the Disposal of Solid Radioactive Waste 0 08 03666 3
:RP Publication 45 (Annals of the ICRP Vol. 15 No. 3, 1986)
uantitative Bases for Developing a Unified Index of Harm 0 08 033665 5

:RP Publication 44 (Annals of the ICRP Vol. 15 No. 2, 1985)
otection of the Patient in Radiation Therapy 0 08 032336 7
:RP Publication 43 (Annals of the ICRP Vol. 15 No. 1, 1984)
inciples of Monitoring for the Radiation Protection of the Population 0 08 032335 9
:RP Publication 41 (Annals of the ICRP Vol. 14 No. 3, 1984)
onstochastic Effects of Ionizing Radiation 0 08 032333 2
:RP Publication 40 (Annals of the ICRP Vol. 14 No. 2, 1984)
otection of the Public in the Event of Major Radiation Accidents:
inciples for Planning 0 08 032302 2
:RP Publication 39 (Annals of the ICRP Vol. 14 No. 1, 1984)
inciples for Limiting Exposure of the Public to Natural Sources of
adiation 0 08 031503 8
:RP Publication 38 (Annals of the ICRP Vols. 11–13, 1983)
adionuclide Transformations: Energy and Intensity of Emissions 0 08 030761 2
:RP Publication 37 (Annals of the ICRP Vol. 10 No. 2/3, 1983)
ost-Benefit Analysis in the Optimization of Radiation Protection 0 08 029817 6

Annals of the ICRP

Aims and Scope

The International Commission on Radiological Protection was founded in 1928 to advan for the public benefit the science of radiological protection. The ICRP provid recommendations and guidance on protection against the risks associated with ionisi radiation, from artificial sources as widely used in medicine, general industry and nucle enterprises, and from naturally occurring sources. These reports and recommendations a published four times each year on behalf of the ICRP as the journal *Annals of the ICR* Each issue provides in-depth coverage of a specific subject area.

Subscribers to the journal receive each new report as soon as it appears so that they are ke up to date on the latest developments in this important field. While many subscribers pre to acquire a complete set of ICRP reports and recommendations, single issues of the journ are also available separately for those individuals and organizations needing a single repc covering their own field of interest. Please order through your bookseller, subscription age or direct from the publisher.

Future publications of the ICRP

(Please note that these reports may be subject to late changes and alterations)

ICRP Publication –, *Age-dependent Doses to Members of the Public from Intake of Radionuclides. Part Embryo and Foetus* (1998/99).

ICRP Publication –, *Basic Anatomical and Physiological Parameters for Use in Radiological Protecti Part 2. Anatomy, Physiology, and Elemental Composition* (2000).

ICRP Publication –, *Dosimetric Model for the Gastro-Intestinal Tract* (2000/2001).

Supporting Guidance Material

The ICRP Database of Dose Coefficients: Workers and Members of the Public (1998, on CD-ROM).

Use of the Respiratory Tract Model for Calculating Doses for Specified Inhaled Chemical Forms Radionuclides (1999).

The ICRP Database of Dose Coefficients: Embryo and Foetus (1999, on CD-ROM).

0146-6453(2000)30:1;1-A

In compliance with GPSR, should you have any concern about the safety of this product, please advise: International Associates Auditing and Certification Limited, The Black Church, St Mary's Place, Dublin 7, D07 P4AX Ireland EUAR@ie.ia-net.com

Printed and bound by CPI Group (UK) Ltd, Croydon, CR0 4YY
14/09/2026
02216204-0002